Unique Animals and Birds of Australia

Michael Morecombe

Unique Animals and Birds of Australia

Michael Morecombe

This edition published in 1990 by PR Books,
5 Skyline Place, Frenchs Forest, NSW, Australia, 2086
Telephone (02) 975 1700, facsimile (02) 975 1711
First published in 1973 by Lansdowne Press Pty Ltd

Typeset by Trade Composition Pty Ltd
Printed in Singapore by Toppan Printing Co. (S) Pte Ltd

National Library of Australia Cataloguing-in-Publication data

Morcombe, Michael.
Unique animals & birds.

New ed.
Includes index.
ISBN 1 875113 26 6.

1. Zoology—Australia. 2. Zoology—Australia—Pictorial works. I. Title.

591.994.

Front Jacket; main photograph: Frilled Lizard *Chlamydosaurus kingi* Probably the most spectacular and colourful of all Australian lizards, the Frilled has developed to the full the survival tactics of surprise and bluff. The abruptness of the display is all important. At one instant the lizard is running—and it is very fast, dashing along upon its hind legs—then suddenly it has turned on its foe and explodes into a shape that, from the front, is now at least four times as big. The lizard displays warning colours of red and orange, and a gaping yellow mouth.

The Frilled Lizard has a wide distribution across northern Australia. The colours of the frill membrane vary considerably, being usually yellow with black and white markings in central Queensland, and orange with reddish, black and white markings in the Northern Territory and the Kimberleys of Western Australia.

Front Jacket; inset: Orange Chat
Back Jacket: New Holland Honeyeater or Yellow-winged Honeyeater
Half Title Page: Western Rosella
Title Page: (right) Sugar Glider and (left) Honey Possum
Below: Echidna or Spiny Anteater

CONTENTS

Rabbit-eared Bandicoot

Introduction

It has long been recognised that Australia has a remarkable number of 'unique' animals. This island continent shelters a whole host of marsupial species found nowhere else on earth; with New Guinea it shares the distinction of being the only place on earth where the three basic types of mammal — monotreme (egg-laying), marsupial (pouched) and placental (womb-bearing) — exist side by side.

Australia's bird life, although not as strikingly different, includes many species with distinctive plumage and unique patterns of behavior.

Australia has become a fauna region truly unique because it has been cut off by oceans from all other land masses for a tremendous length of time. Early primitive marsupials which were marooned on this continent found it a sanctuary, free of competition from the more advanced placental mammals which were appearing on other continents. Eventually marsupials vanished from most other parts of the globe, replaced by the proliferating placental mammals, leaving isolated Australia almost a living museum of early forms of mammal life.

But animals such as the kangaroo and possum have not always been unique to Australia. In distant past ages Europe had animals which shared many of their characteristics, and there are still species of opossum and other marsupials to be found in North and South America.

When continents are isolated from one another by vast expanses of ocean, the animal populations on each land mass evolve their own particular characteristics, giving rise to new species, in a way which can be explained by the Darwinian theories of natural selection.

There have, of course, been several puzzles in the present-day distribution of mammals. Why should two continents as far apart as Australia and South America have almost all of the world's marsupials? And why did Australia not have any of the placental mammals until comparatively recently in its history?

The answers probably lie in an old theory which in recent years has become more and more credible, until now it has become an established fact. This is the theory of continental drift, which suggests that the continents and islands of the world do not have a fixed place on the surface of the globe but, together with, and as part of, large slabs or plates of the ocean floor, are drifting. The movement is extremely slow, but even 25 mm a year adds up to some 19,000 km in a hundred million years, which is not a long time in geological terms.

At times in the distant past the plates carrying continents have collided (pushing up mountain ranges such as the Himalayas) allowing the intermingling of their faunas.

Zoologists have worked out a 'timetable' comparing the estimated positions of continents in the distant past with fossils of the period. Thus the fossil record of the time when Australia was calculated to have been in contact with South America via the Antarctic continent, some forty-three millions years ago, shows significant similarities in animal life.

Australia and New Guinea then separated from Antarctica, and drifted together northwards, while South America on the other side of the globe was also drifting.

From this time onwards the marsupials inhabiting each continent took separate lines of evolution according to the conditions prevailing. The Antarctic continent became a region of extreme cold, rendering extinct most forms of life. South America joined with North America, allowing a flood of the more advanced placental mammals to invade the south, competing with the marsupials, causing the extinction of many species, and reducing the marsupial population to its present minor part in the total mammal fauna of that continent.

In comparatively recent times a single species of South American marsupial, the Opossum, has colonised North America, and become well established.

Australia likewise drifted for thirty million years or more completely isolated from any other land mass.

During this time its animal life was a community completely cut off from outside influences. The marsupials evolved towards their present multiplicity and diversity of species, and (apart from the few monotremes) probably made up the entire mammal fauna.

Eventually, some ten million years ago, this huge drifting plate of the earth's crust carrying Australia and New Guinea approached the Indonesian islands, which lie along the submerged southern edge of a plate which, to the north, rises again above the sea as the Asian land mass.

At this time the mammals of Asia began to find their way into Australia. At first came those that could cross the intervening sea, the bats and flying foxes, then rodents probably coming on windblown driftwood. Finally came Aboriginal man, bringing a type of dog which, spreading and running wild during the following thousands of years, became the distinctive, reddish-brown Dingo that is now found only in Australia.

Although most of the 650 species of birds are unique to Australia they do not stand out from the rest of the world fauna to the same degree as do the mammals.

In Australia is a whole host of brilliantly plumaged parrots and cockatoos, finches, honeyeaters, wrens and robins; but it is probably for those birds that are unique in the peculiarities of their behavior that Australia is best known. Here are found the megapodes — the Mallee Fowl, Brush Turkey and Jungle Fowl — as well as bowerbirds, lyrebirds, and the world's second largest bird, the Emu.

Australia has many species of fairy wrens, brightly coloured little birds generally with blue dominant in the plumage. Their long tails are carried upright when they are perched or hopping on the ground, but trail behind like long streamers when they fly. This Blue-and-white Wren occurs in the drier parts, generally inland areas, of all mainland States.

Right:

The Koala, one of Australia's best-known marsupials, is unique to the forests of eastern Australia.

Emu

Travelling fast, a female Bennett's Wallaby and her large joey leap a tree trunk. The large herbivores of other continents have a running or galloping gait, but most of Australia's herbivorous marsupials, those that live on the ground, have developed two-footed leaping as a means of fast travel. The erratic bounding action may give greater speed through forest undergrowth, enable the Wallaby to change direction suddenly, dodge around or bound over bushes, logs, stumps and boulders.

1 The Unique Marsupials

The first impressions of Australian marsupials were those recorded by the Dutch navigators who discovered parts of the western and northern coastline during the seventeenth century. Groping for adequate comparisons, these first eyewitnesses tried to describe animals which had no likeness elsewhere on earth. No word yet existed which could possibly summarise an animal as different from the familiar deer, horses or cattle as the kangaroos they saw when first coming ashore on this remote continent.

Writing for those back home, the navigators tried to create an image of the appearance of these new animals by likening them to rat, cat or jerboa, then detailing the differences. Even today we have no adequate common names for many of our unique mammals, and continue to rely upon names borrowed from totally unrelated creatures. The large spotted carnivorous marsupial that hunts in the tree tops of eastern Australia has no kinship with either tiger or cat; the striped marsupial carnivore that inhabits Tasmania is quite unrelated to wolf or tiger, while the small insectivorous marsupials have very little in common with ordinary mice other than size.

Pelsaert, a Dutch navigator wrecked on the Houtman Abrolhos in 1629, wrote of the little Tammar wallabies that inhabit those low scrubby sandy islands:

> We found on these islands large numbers of a species of cats, which are very strange creatures; they are about the size of a hare, their head resembling the head of a civet-cat; the forepaws are very short, about the length of a finger, on which the animal has five small nails or fingers, resembling those of a monkey's forepaw. Its two hind legs on the contrary, are upwards of half an ell in length, and it walks on these only, on the flat of the heavy part of the leg, so that it does not run fast. Its tail is very long, like that of a long-tailed monkey; if it eats, it sits on its hind-legs, and clutches its food with its forepaws, just like a squirrel or monkey . . .

Willem de Vlamingh, landing on Rottnest Island in 1696, saw the little rat-kangaroos which are still common on that island. This species, the Quokka, he wrote of as being:

> a kind of rat as big as a common cat.

The first English captain to visit Australia, William Dampier, saw on Dirk Hartog Island the Banded Hare-wallaby:

> The Land-Animals that we saw here were only a Sort of Raccoons, different from those of the West Indies, chiefly as to their Legs; for these have very short Fore-Legs; but go jumping upon them as the others do . . .

These early sightings on Australia's western coast were all small wallabies. Large kangaroos were described in Cook's journal in June 1770:

> One of the men saw an animal something less than a grey hound, . . I saw myself this morning a little way from the ship one of the Animals before spoke of, it was of a light Mouse colour and the full size of a grey hound and shaped in every respect like one, with a long tail which it carried like a grey hound, in short I should have taken it for a wild dog but for its walking or running in which it jumped like a Hare or a dear [sic]; Another of them was seen today by some of our people who saw the first, they describe them as having very small legs and the print of the foot like that of a goat, but this I could not see myself because the ground I saw it upon was too hard and the length of the grass hindered my seeing its legs.

Later, one of Cook's men shot a kangaroo, and this gave the Englishmen an opportunity to examine more closely one of these unique animals.

> Mr. Gore being out in the country shott one of the Animals before spoke of, it was a small one of the sort weighing only 28 pound clear of the entrails. The head neck and shoulders of this Animal was very small in proportion to the other part; the tail was nearly as long as the body, thick next the rump and tapering towards the end; the fore legs were 8 Inch long and the hind 22, its progression is by hoping or jumping 7 or 8 feet at each hop upon its hind legs only, for in this it makes no use of the fore, which seem to be designed for scratching in the ground & C^a^. The Skin is cover'd with a short hairy fur of a dark Mouse or Grey Colour. Excepting the head and ears which I thought was something like a Hare's, it bears no sort of resemblance to any European Animal I ever saw; it is said to bear much resemblance to the Gerbua excepting in size, the Gerbua being no larger than a common rat.

The largest of Australia's unique marsupials is the Red Kangaroo. Boomers, the big old males, may weigh more than 82 kg, yet at birth the joey is only 18.5mm in length.

Kangaroos 'sparring'

These early records of the unique Australian marsupials emphasise the strangeness of their outward appearance; the first hint of a difference of more fundamental nature was Pelsaert's observation of tiny young in the pouches of Tammar Wallabies in 1629:

The manner of generation or procreation is exceedingly strange and highly worth observing. Below the belly the female carries a pouch, into which you may put your hand; inside this pouch are her nipples, and we have found that the young ones grow up in this pouch with the nipples in their mouths. We have seen some young ones lying there which were only the size of a bean, though at the same time perfectly proportioned, so that it seems certain that they grow there out of the nipples of the mammae, from which they draw their food, until they grow up, and are able to walk.

The pouch quickly became the distinctive feature of Australian marsupials for all who saw or read about these unique animals. Because the pouch young are at first only a few millimetres in length it was widely held that they were actually born in the pouch, as tiny 'buds' at the tips of the teats. This fallacy began with Pelsaert's observation in 1629, and persists to some degree to this day.

It is of course anatomically impossible for the marsupial young to be born on the teats, which are solely for the provision of milk, as with other mammals. The young is in fact born in the same manner as any other mammal, although extremely small at birth. There are however certain characteristic differences in the internal reproductive organs of the marsupial female.

Early scientists, who saw embryonic young attached to the teats in the pouches of kangaroos, knew from their dissections that the young marsupial was conceived in the usual mammal fashion. They assumed the newborn young must surely be placed in the pouch by its mother, or at least given some assistance. But not only do the incredibly small embryonic young struggle through the fur to find the pouch opening, but once inside find a teat and attach themselves firmly to it.

The first Australian observation of the journey of an embryonic kangaroo (a newborn Tammar Wallaby) to its mother's pouch was recorded in 1830 by Alexander Collie, the surgeon aboard H.M. sloop *Sulphur*. The ship was at the time anchored in Cockburn Sound, off the present port of Fremantle.

An officer of H.M.S. Success at present here, observed a Kangaroo in the act of parturition. When the foetus was expelled . . . the mother was lying partly on one side . . . resting against the side of the cage where she was confined . . . and the very diminutive young, when brought forth, crept among the fur of the mother towards her belly and towards the opening of the abdominal pouch; whilst she, with her head turned towards her tender offspring, seemed to watch its progress, which was about as expeditious as a snail.

The young marsupial receives no assistance from its mother in its struggle through the fur to the pouch. It is endowed with all the necessary adaptations at birth, having well-developed and active forelimbs, jaw muscles and tongue. Although the eyes and ears are not yet functioning, it has large nostrils, and presumably a well-developed sense of smell which probably guides it to the pouch.

It is thought that marsupials entered Australia around eighty million years ago, possibly from South America via Antarctica, at a time when the Antarctic climate was much warmer and these land masses much closer together. South America today still has marsupials among its mammal fauna, but the fossil record shows that it had a far greater number in the past. An alternative hypothesis holds that they entered Australia prior to that continent's separation from Asia, towards the end of the Cretaceous period some sixty million years ago.

Today marsupials survive only in Australia, New Guinea and South America, while North America has a single species, the Opossum, a comparatively recent arrival from South America. Australia now has some seventy-five per cent of known marsupials.

Some indication of the former wide geographical range of marsupials is shown by the discovery of marsupial fossils in Europe and North America.

Little is known of the very early evolution of marsupials in Australia because the fossil record does not extend back beyond the Oligocene period, about thirty million years ago. But by the Pleistocene period, around one million years ago, there was a rich marsupial fauna, including giant forms now extinct. Amongst these was the Diprotodon, a four-footed herbivorous marsupial the size of a rhinoceros; and *Thylacoleo*, a panther-sized predator with unusual shearing teeth and prehensile claws.

The marsupials today are a widely diversified group in appearance and habits, but all retain the features which are characteristic of the marsupial branch of mammal evolution. In spreading across the face of the continent to inhabit all its vastly differing regions, the first marsupials became separated from others of their own kind. Each isolated population gradually aquired characteristics to suit the local environment, and each became eventually, in its own way, different enough to be a distinct new species.

Feather-tailed Glider

In this way, over the millions of years, they have multiplied in species, genera and families to fill almost every available habitat from jungle to desert dunes. By the time of the arrival of European man there were about 120 species of marsupials.

The marsupial carnivores were probably the first Australian marsupials. From these unspecialised primitive mouse and rat-like insect-eaters have evolved not only the larger carnivores, but also the whole multitude of herbivorous marsupials.

Although the vast majority of the marsupials in Australia today are quite small, and prey principally upon insects, there are some quite large predators: The Tiger Cat, the Tasmanian Devil and the Tasmanian Wolf or Tiger

Many of the small insectivorous marsupials such as the Dunnart are easily mistaken for mice or rats, but any similarity is quite superficial. The alert little marsupial-mice have sharp-pointed faces when compared with the blunt-nosed faces of the rodents, and they have white needle-pointed cat-like teeth instead of the yellowish chisel-shaped teeth of ordinary rats and mice. Being marsupials they carry their young in a pouch, which generally has not yet evolved to becoming a deep pocket but is merely a shallow depression which contains the teats where the young remain firmly attached for a long period after their birth.

Settlement and development have greatly thinned out our marsupial carnivores, and brought some species very close to extinction. The Tasmanian Tiger is extremely rare if not extinct, the Tiger Cat is rare in most areas except Tasmania, the Eastern Native-cat is rare on the mainland, the Dibbler or Freckled Marsupial-mouse may be nearly extinct.

The great kangaroos include the Red of the inland plains, the greys of southern forests and woodlands, and the wallaroos or euros which prefer rocky hills of both coastal and inland areas. Compared with the Red Kangaroo and the two species of grey, wallaroos look shorter and more heavily built, an impression resulting partly from their long shaggy fur. Closely related to the wallaroos is the Antilopine Kangaroo of the northern tropical grasslands.

The largest wallabies have a total length, nose to tail-tip of about 92 cm (for the Red-necked Wallaby), but other than size and the marking of individual species there is no obvious difference between large wallabies and the kangaroos.

The wallabies are a very diverse group, having expanded into a bewildering array of species and subspecies that occupy almost every available habitat. Their wide range of sizes probably results from their environment. They inhabit more densely vegetated country than the kangaroos and, the smaller the wallabies, the thicker the undergrowth they choose for their home.

Possums and gliders all live and feed in the tree tops. Being closely related, they have each inherited from some common ancestor certain adaptations for the arboreal way of life

Each has a long tail which is either prehensile (and therefore able to curl around branches and grip more tightly than a hand); or bushy, fluffy or feather-like, which is used as a balance and rudder during long leaps and gliding flights.

Whiptail Wallaby

But the most distinctive feature of the family Phalangeridae to which possums, gliders and cuscuses belong, is the hand-like shape of the hind feet. The clawless big toe, like a human thumb, is opposable to the rest of the foot, giving a firm grasp on the branches. There are strong claws on all other toes of hands and feet. The second and third toes, like those of kangaroos, are combined or syndactylous.

The possums may be considered in three main groups: the brushtails, the ringtails and the pigmy possums. Three species not fitting any of these groups are the Striped Possum, the Leadbeater's Possum and the Noolbenger or Honey Possum.

Australia's remarkable flying phalangers or gliders have evolved on three separate occasions, from three different groups of non-gliding possums. From the ringtails have been developed the Greater and Yellow-bellied Gliders, from the Leadbeater's Possum have come Sugar and Squirrel Gliders, and from the Pigmy possums, the tiny Feathertail Glider.

The monkey-like cuscuses have come to Australia from New Guinea. Cuscuses, together with possums and gliders, belong to the family Phalangeridae, identified by the prehensile tail and hand-like hind foot.

The Koala, although now highly specialised for life in the tree tops, has at some stage of evolution lost the tail which is so useful to arboreal mammals. It is thought that this may indicate a period of terrestrial life somewhere in its ancestry. The Koala appears to be more closely related to the wombats than to other marsupials. For example, it differs from the possums in having cheek pouches which are associated with the chewing of the bulky leaf food; in wombats there are similar though rudimentary cheek pouches. Both Koala and wombats have pouches which open backwards, but possum pouches open forward.

The destruction of forests, epidemics and the killing of thousands for their fur reduced Koala numbers to a dangerously low level.

Australia has four species of wombats, in two genera, the forest wombats, *Vombatus*, and the plain wombats, *Lasiorhinus*. The former genus, to which the Common and Tasmanian Wombats belong, is probably safe, inhabiting forests and rocky ranges over a wide area, but the various plain wombats (the species of hairy-nosed wombats) are mostly confined to restricted localities where they are very vulnerable to interference and therefore possible extinction.

Wallabies are distinguished from kangaroos only by their smaller size, the largest of the wallabies having a nose to tail-tip length of no more than 2 m. This species, the Agile Wallaby, inhabits the lush tropical grasslands of northern Australia, from the Kimberleys to northern Queensland.

Right:

As the female Tammar Wallaby sits up alert her large joey (too big to ride in the pouch) thrusts its head in for a drink. As soon as a joey leaves the pouch for the last time, another is born, finds its way through the fur, and attaches to a nipple. That second joey, while still an embryo in the womb, could be held in a state of suspended development, resuming its growth towards birth when the pouch becomes vacant. Thus a kangaroo always has a reserve young ready to move into the pouch, even if the first joey should die and leave the pouch empty sooner than would otherwise be the case.

Below:

Almost large enough to fend for itself, this Bennett's Wallaby joey, sharing a leafy twig with its mother, has already been evicted from the pouch. The Bennett's Wallaby occurs in Tasmania and on King Island in Bass Strait; a very similar sub-species, the Red-necked Wallaby, inhabits parts of the Australian mainland.

Right:

As its mother basks in the sun a restless joey, almost too big for the pouch, watches the events of the outside world. He would have started leaving the pouch at seven months, will make his final exit at about nine months, but will continue to suckle from outside until about one year old.

Below:

Young possums that have outgrown the pouch are carried on their mother's back until almost full-grown. Ringtails usually have a 'litter' of two, but the mortality rate is very high, forty-five to seventy-five per cent dying before they become reproductively mature at the age of eighteen months. Ringtails build nests in the trees, large, football-shaped masses of leaves and twigs hidden in dense foliage. Here the mother and her young sleep the days away, emerging at dusk to feed on leaves and native fruits.

As with all other marsupials, Numbats are born incredibly small and must climb through the fur to the pouch, which is not the deep pocket of kangaroos, but merely a shallow depression. The three or four baby Numbats cling with mouths and paws, rarely moving and never releasing their tight grip until much larger. At the stage shown here they are about 38mm in length, and their stripes are faintly visible.

Right:

One of the most attractive of the large possums, the Bobuck or Short-eared Possum has an alert, almost surprised expression, big bulging brown eyes almost like those of the ringtails, stubby little ears and luxuriant silvery fur. His ears are only about half the length of the ears of other brushtails, although he is as big as the common brushtails. His hind feet, as well as the front, are hand-like, a contrast to the narrow, elongated hind feet of kangaroos. With a first toe like a thumb, opposable, the possum's hind foot is capable of grasping.

Below:

The Common Brushtail is very widespread in forested coastal areas of Australia. As would be expected with a distribution ranging from tropical rainforests to cold southern mountains, this species shows great variation in colour and other features. There are about eleven different sub-species, some pale grey, some brown, and some black in colour.

Left:

Gliding down from a higher branch a mouse-sized Feathertail Glider at the last instant swoops upwards, slowing its speed before dropping on to the cluster of flowers where it will hunt for insects and nectar.

Below Left:

A Greater Glider, holding its 30 cm wide parachute-shaped body almost vertically to check its speed, lands at the base of a tree after gliding down from the top of a tall eucalypt some distance away. The opening of her pouch is visible on the soft white undersurface fur.

Below:

With fur of a mottled mossy greenish-brown colour, a Green Ringtail can be almost invisible when high in the foliage of a north Queensland jungle tree. This possum is unique to the mountain rainforests of the Atherton Tableland area. Like other ringtails, it carries its young on the lower part of its back.

2 Numbat, Bandicoot and Wombat

The Numbat, the bandicoots and the wombats are all ground-dwelling marsupials, but apart from this have little in common. Each is, however, well adapted to a particular way of life, illustrating the versatility of the marsupials in occupying almost every available niche in the Australian environment.

A unique, highly-specialised carnivore which feeds almost entirely upon termites, the Numbat has a long, rubber-like extensile tongue and other features facilitating the finding and capture of this food.

Of the wombats, Australia has four species, in two groups — the forest wombats and the plain wombats. Although there is evidence that wombats may once have been tree-dwellers, sharing some distant ancestry with possums and the Koala, they have since then had a very long history as terrestrial animals, becoming so specialised in their present way of life that they are now placed in a family of their own.

Bandicoots, of which there are many species, are mixed feeders, that is they eat insects, vegetable matter, occasionally small reptiles, and even very small mammals. They have managed to colonise most parts of Australia, there being desert-dwelling species, and others that inhabit the wettest coastal rainforest.

Numbats, probably the most beautifully coloured of all Australian mammals, once occurred as far eastwards as the western edge of New South Wales, but have long been extinct in the eastern half of the continent. They are now restricted to southern Western Australia and north-western parts of South Australia.

The Marsupial Anteater (a common alternative name for the Numbat) is about 23 to 30 cm in length of head-and-body, plus another 17 to 20 cm for the brushy tail; it usually weighs about half a kilogram. The general colour is grey-brown to reddish-brown, with prominent white and black stripes across the back and rump. The face is very sharply pointed and the long slender tongue can be extended 10 cm. The teeth are extremely small, degenerate, and vary in number from fifty to fifty-two, the largest number found in any land mammal.

Unlike most wild animals, the Numbat is out and about during daylight hours. It may be glimpsed dashing across a bush track, tail arched high above its back, or be seen scratching around a termite mound for the 'white ants' upon which it feeds almost exclusively.

This defenceless creature has survived the introduction of the fox because this predator is nocturnal. When feeding the Numbat is constantly alert for the slightest sound of danger. If alarmed it will sit upright, or stand on hind legs, then go dashing away through the undergrowth, with bushy tail arched over its back, to the safety of one or other of the hollow logs scattered through its territory. Once wedged deep inside the narrow cavity of a long heavy log the Numbat is safe from most of its enemies.

While foxes appear to have had little effect upon the Numbats of south-western Australia, natural predators probably take an occasional animal. The Wedge-tail Eagle, Australian Little Eagle, Goshawk and Carpet Snake all occur in the region, and all would be capable of taking an unwary Numbat.

Extensive studies of the Numbat in the wild have been made by J.H. Calaby of the Wildlife Survey Section of the CSIRO. Previous observations by pioneer naturalist John Gilbert, published by John Gould in 1845, were made within a few years of the discovery of the Numbat, and for a long time remained the only information available on this marsupial.

In his study of the Numbat in the south-western corner of Western Australia, Calaby found that while its range has shrunk very considerably over the past thirty years, it has not suffered the catastrophic reduction in numbers or the near-extinction that has befallen many other small native mammals.

There are two sub-species, the south-western animal, *Myrmecobius fasciatus fasciatus*, found scattered in remnants of bush over an area between Perth, Kalgoorlie and Albany; and the Rusty Numbat,

Unlike most of Australia's small native mammals the Numbat, or Marsupial Anteater, is out and about during daylight hours.

M. f. rufus, which occurs in north-western parts of South Australia and adjacent border areas of Western Australia.

Calaby's study of the Numbat was done in a Western Australian State Forest known as Dryandra, in the Narrogin district, where a flourishing population of these anteaters was known to occur. This forest, about thirty-nine square kilometres in area, has been managed by the Forest Department in such a way as to maintain a suitable habitat and sanctuary for the Numbat for many years.

The dominant trees at Dryandra are the Wandoo and the Powder-bark Wandoo, which with others of similar characteristics are the vital element of the Numbat's environment.

These trees are readily attacked by termites, and develop many hollows in limbs and trunks. The ground has become littered with fallen trees, limbs and branches, mostly hollow. The hollow logs are the indispensable shelters for the Numbat, while the termite responsible is also the anteater's chief food.

In the forest the ground is strewn with sticks, chips and dead leaves, most of which, if lifted, expose termites. Here and there are the large clay mounds of termite nests.

Much of the forest has an understorey of low, rather prickly scrub, including species of *Dryandra*. This provides a dense 'roof' of foliage a metre to one and a half metres above the ground, yet relatively open at ground level. This would permit the little Numbat to run swiftly towards the nearest hollow log, while blocking the dive of eagle or goshawk or hindering the passage of any larger animal such as a fox.

Only where the Wandoo woodland has maintained its natural features of abundant hollow logs, together with a low open shrub stratum, can the Numbat survive. It has vanished from far too many blocks of bushland through the Western Australian wheatbelt country not because anyone would kill a Numbat, but because frequent burning off has replaced the protective shrubs and logs with a sea of grass that is useless as a native mammal habitat, offering no protection from foxes, cats and other predators.

Termites are very much the favoured food of the Numbat. Although many small ants are eaten, this seems mainly accidental, occurring when the Numbat breaks into ant chambers which often occur in parts of a termite nest. The termites and ants are swallowed without being chewed — the small, soft-bodied insects hardly require mastication. Because so much sharply angular-grained sand is collected on the Numbat's long sticky tongue along with the termites, regular chewing would cause considerable tooth wear.

As the Numbat is a marsupial its young are born extremely small. The female has no pouch and the young are firmly attached to the teat by their mouths and also hang on tightly by their hands to the long coarse hair of the 'pouch area'. Their mother does

Numbat

nothing that shows she is aware of their presence. She pays no attention to them, and in climbing over branches and pushing into narrow hollow logs scrapes them against rough wood. If they are touched their already curled little bodies curl tighter, and their tiny hands grip the fur even more strongly.

In south-western Australia the Numbat has a definite breeding season, the young being born during summer, autumn and early winter. They are carried by their mother through the winter, but by late spring some may be seen foraging on their own; there are four in the normal litter.

The 'pouch life', the period for which the young cling beneath their mother, is believed to be between four and six weeks, which is a long time compared with other marsupials of similar size. By the time they are a little more than 2.5 to 3.0 cm in length they are covered with a thin golden downy fur; young 5 cm in length are beginning to show the dark bars across the rump.

When they become larger the young are detached from the teats and live in a nest in a hollow log, or perhaps occasionally in a hole in the ground.

The principal danger to the survival of this very beautiful species of native mammal is the clearing of land for farming. It can exist quite well on blocks of bushland left between cleared paddocks, provided that the ground strata of shrubs and logs remains. The future of the species can be assured by the provision of adequate reserves or national parks, which must be large as the Numbat has a low population density, and the minimum number of animals required to maintain a breeding colony is at present unknown.

Wombats have a wide distribution in southern Australia, from southern Queensland and through New South Wales, Victoria, Tasmania and South Australia to the border of Western Australia.

These large, solidly built animals are entirely vegetarian, and have distinctive rodent-like teeth unlike those of any other marsupial. There is only a single pair of upper and lower incisors which, like

The Tasmanian Wombats belong to the forest or naked-nosed genus, and are the smallest of the wombats. They are now thought to be insular forms of the Common Wombat rather than distinct species.

the incisors of rodents, grow continuously and are worn away and sharpened in use.

All wombats are powerful diggers, their stocky build and shovel-like claws enabling them to excavate burrows about a metre in diameter and up to 20 m in length.

Australia has four species of wombats, grouped in two genera: the forest wombats, *Vombatus*, and the plain wombats, *Lasiorhinus*. The former genus, to which the Common Wombat and the Tasmanian Wombat belong, inhabit forests and rocky ranges over a wide area. These are probably safe in the foreseeable future, but some of the plain wombats (the various hairy-nosed wombats) are confined to small areas where they are very vulnerable to interference, and could become extinct through loss of habitat.

Forest wombats (the Common Wombat, the Tasmanian Wombat and the Island Wombat from Bass Strait) are the best known. They can be recognised by their coarse thick body fur, short ears, and naked snouts.

The plains wombats have noses of a rather different shape, covered with fine hair. Their ears are much longer and pointed, and their body fur is soft and silky. When fully grown, a wombat is about a metre in length, approximately 30 cm high, and weighs 27 to 32 kg. It has short but extremely powerful legs, with strong claws well adapted for digging. The silky fur is usually a beautiful silvery-grey, scattered with longer black guard-hairs. The tail is a stumpy 25 mm appendage.

Studies have been made of wombats, particularly the endangered Hairy-nosed Wombat. These animals have become extremely specialised in their feeding habits. They prefer one particular type of perennial grass, but will eat other grasses when necessary, particularly in times of drought; unlike the Common Wombat, the Hairy-nosed will not eat roots, bark or leaves.

Even under drought conditions Hairy-nosed Wombats do not appear to drink at stock watering places, but when there is natural rainfall they come out of their burrows to drink from pools that collect in hollows of the limestone capping which covers much of the region. It is said that on such occasions they can be approached to within about a metre.

In times of drought it seems they go without water for three or four months or more. This ability of the wombats to live without water derives partly from their habit of living in deep burrows. In hot weather they will rarely be seen above ground during the day; instead they sleep deep down in their cool, humid burrows, conserving their energy to avoid loss of water.

During mild sunny weather, autumn or spring, wombats may often be seen sitting or lying lazily on the mounds of soil around their burrow entrances.

The burrows of the Hairy-nosed Wombat are often clustered together to form extensive warrens, with the entrances meeting in a large pit. Most burrows are dug down under the limestone capping, going 3 m or more below the surface. There are also shallow earth burrows which appear to be temporary homes for young animals when they leave their mothers.

Wombats, like most other marsupials, have pouches, but unlike the well-known kangaroo pouch, the wombat pouch opens more towards the rear, perhaps because it is less likely to fill with sand during tunelling operations. Its opening is kept closed by a strong circular muscle to prevent the baby falling out.

The young wombat is a typical marsupial in that it is born very immature and extremely small — only about 2.5 cm in length. After birth it crawls through the fur to its mother's pouch, fastens to one of the two teats, and continues its growth in the pouch for several months before venturing again into the outside world.

Of bandicoots, there are many species, but all share several main characteristics: they possess the many-incisored or polyprotodent dentition of the marsupial carnivores and they have a foot structure similar to that of the vegetarian kangaroos and possums.

The bandicoots on average are mixed feeders, eating insects, vegetable matter, occasional small mammals and reptiles; their food preferences vary from one species to another, some being entirely insectivorous, and some entirely vegetarian.

Many of the bandicoots are grouped according to the relative lengths of their noses. All bandicoots have long slender snouts. The members of one genus

Right:

The Numbat, probably the most beautifully coloured of all of Australia's furred animals, may be glimpsed scratching around termite mounds or turning over leaves and twigs in its constant search for the 'white ants' which are its principal food. At the slightest sound it will dash for one of the many hollow logs in its territory.

have quite long snouts, but despite this are known as 'short-nosed bandicoots' because their noses are shorter than other extremely long-snouted bandicoots.

One of the commonest, most widely distributed and best known of all the short-nosed bandicoots is the species *Isoodon obesulus*, the Quenda, or Brown Bandicoot, or Southern Short-nosed Bandicoot. This animal is about the size of a rabbit, head and body together 35 cm, and the tail a rather short 12 cm. The muzzle is pointed, ears short and rounded, and the fur short, coarse, and grizzled yellowish-brown, white on the belly.

The Quenda occurs in eastern Australia from Cape York Peninsula through New South Wales to southern Victoria and south-eastern parts of South Australia; it is found also in Tasmania and Western Australia.

Inhabiting forests and heathlands where ever there is dense ground cover, the Quenda is nocturnal and insectivorous, leaving little conical pits where ever it has been digging for beetles and insect larvae. This bandicoot constructs a rough nest of sticks on the ground under dense vegetation; there is usually no well defined entrance, as it just pushes under the rough pile of loose debris. Seven young form the normal litter.

Other short-nosed bandicoots, of more limited distribution, are the Brindled Bandicoot, and the Golden Bandicoot.

There are two genera of long-nosed Bandicoots, the genus *Perameles* containing many common species. These animals are nocturnal, insectivorous, and have incredibly long, thin, almost tubular snouts. Compared with the short-nosed bandicoots, they have longer legs and ears, and look more lightly built.

There are, in addition to the short-nosed and long-nosed bandicoots, two genera that are especially interesting — the Rabbit-eared bandicoots, and the Pig-footed Bandicoot.

The two species of Rabbit-eared bandicoots, both now very rare, are among the most beautiful and graceful of all native mammals, having large rabbit-

Long-nosed Bandicoot

like ears, and long silky black-and-white or all-white tails. They are powerful diggers, and always live in deep burrows. Rabbit-eared bandicoots are carnivorous, feeding principally upon termites, beetle larvae and other insects.

One species, the Common Rabbit-eared Bandicoot, Dalgyte or Bilby, formerly occurred in Western Australia from the Kimberleys to the south-west, in western and northern parts of South Australia, western New South Wales and far south-western Queensland. The Lesser Rabbit-eared Bandicoot, or Yallara, is confined to sandhill country of the Simpson Desert.

The Pig-footed Bandicoot is now extremely rare, and may be extinct, the last specimen being collected in 1907. This unusual animal, somewhat smaller than a rabbit, looked rather like a miniature deer, with long ears, crested tail, and long thin legs. It inhabited the arid and semi-arid woodlands, mallee scrub and grasslands of Central Australia, inland Western Australia, South Australia, south-western New South Wales, and northern Western Australia. But it is no longer to be found in any of these places, having vanished, perhaps from the face of the earth, as its native haunts were taken over and greatly changed by sheep, cattle and rabbits.

Below:

Bandicoots are one of the three main branches of marsupial evolution occurring within Australia. They have become adapted to a life of digging, searching for insects and other small invertebrates in the soil, among leaf-litter and around roots. The Quenda or Brown Bandicoot belongs to the short-nosed group of bandicoots — its nose, though long, is far shorter than the noses of some other bandicoots. Like most species the Quenda has harsh, almost spiny fur. These nocturnal animals dig little conical pits in their searching, and sometimes inhabit suburban gardens.

Left:

The Hairy-nosed Wombats of the Nullarbor Plains live in large colonies where their many burrows form a complex of tunnels often with the entrances meeting in a large crater. Although these wombats are most active at dusk and dawn, they often emerge from their burrows in mild calm weather to lie in the sun.

Below:

Lumbering from its burrow a Hairy-nosed Wombat of the Blanchetown area of South Australia shows its distinctive snout. This animal and the Nullarbor wombat are two slightly different geographical forms of the one species of Hairy-nosed Wombat; they can easily be distinguished from the Common Wombat and Tasmanian Wombat not only by the different shape of their muzzle, but also by their soft, fine, silky fur and longer ears.

3 Tasmanian Devil, Tiger Cat and other Carnivores

The fierce-looking black Tasmanian Devil and the spotted Tiger Cat are two of the largest carnivorous marsupials in Australia today, surpassed in size only by the Tasmanian Wolf, which may now be extinct. At the other end of the scale is a whole host of smaller native-cats and insectivorous marsupial-mice, including the world's smallest marsupial, the Planigale.

Carnivorous marsupials were probably the original mammals which entered Australia before this continent broke away from the other land masses; they are regarded by zoologists as the stem forms from which other kinds of Australian marsupials evolved.

Although all are flesh-eating predators, living, according to size, upon mammals, birds, reptiles or insects, they are extremely diverse in habits. Some are arboreal, but most are terrestrial.

The popular easy definition of a marsupial as being a pouched mammal breaks down somewhat in the case of the carnivores, because the pouches of most species are not deep pockets, but are usually no more than shallow depressions which contain the teats and the small young.

The Tasmanian Devil, entirely black but for a few small white markings, is probably the ugliest of all marsupials. Its voice, an eerie whining snarl, is partly responsible for the name 'devil'. About the size of a small dog, the Devil has a short, heavy-set body, short powerful legs, a disproportionately large head, and short tail.

This is one of the few marsupials that does not have an elongated muzzle; its head is rounded, broad and heavily built. The upper jaw projects over the lower, and both head and jaws are extremely powerfully-muscled in keeping with its carnivorous and scavenging habits. The jaws, like those of the hyaena, are adapted for crushing bone as well as for shearing and tearing off flesh, and the teeth make an impressive armament.

Both skull and lower jaw are massively built. The powerful jaw muscles are attached to well-developed ridges at the rear of the lower jaw, and to a prominent crest of bone at the rear upper surface of the skull. The degree of development of these muscle anchor-points gives some indication of the tremendous strength of this carnivorous marsupial.

The Devil runs in a peculiar stiff-bodied canter, almost continuously nosing the ground. The sense of smell is extremely well-developed and appears to be its main guide in hunting.

The preferred habitat of the Tasmanian Devil is the densely forested and often mountainous regions of the island, where its well-worn pads through undergrowth are usually found in the vicinity of streams.

Almost anything the Devil can kill, together with carrion, makes up the diet — snakes, small mammals, lizards, frogs and insects. There are records of it eating fish and crustacea. It eats almost every part of a carcass, bones and all, except for a few small parts such as jaw bones which carry teeth.

Tasmanian Devils are mainly nocturnal, and in their prowlings at times fight noisily among themselves. One naturalist, camped in Devil territory, wrote of 'nights broke by rasping snarls rising to a crescendo of screams, louder and more startling than those of any other bush animal'.

Their lair may be a cave or crevice beneath boulders, among the roots of a fallen tree, or in a hollow log. Usually a bed of bark, leaves and grass is constructed.

Four young are born in May or early June, and quickly find their way into their mother's pouch. They are not seen again for some fifteen weeks, when they first release their grip on the teats and begin to leave the pouch for short periods. They are then well furred and have their eyes open. From this age onwards the young Devils are left in the lair while the female goes hunting; they are finally weaned at five to six months of age.

Equipped with long claws and jaws that open tremendously wide, the Tiger Cat is a powerful predator, living mostly upon birds and small mammals. It is unique to Australia, and the largest marsupial carnivore inhabiting the mainland, having a total length of about 1.2m.

Red-tailed Phascogale

Although Tasmanian Devils do not burrow, their claws are long, aiding them in climbing trees. The young can climb quite well, but the bulky short-legged adults can usually manage to scramble up only those trees which have a sloping trunk.

Tasmanian Devils once lived in mainland Australia, but appear to have been displaced there by the Dingo, which proved to be a more efficient hunter and scavanger. The Devils co-existed with the Tasmanian Wolf, cleaning up remnants of the meal rather than competing for a fresh kill of the same sort of prey.

The Tiger Cat, although hardly the equal of a true tiger, is the largest arboreal marsupial carnivore, growing to a length, nose to tail-tip, of about 1.2 m. Like the Tasmanian Wolf, the Tiger Cat has jaws that open exceptionally wide; it has a reputation, dating back to the earliest days of settlement, for ferocity and stubbornness

The Tiger Cat is principally an animal of the tree tops, occurring in greatest numbers in the heavily forested coastal mountains of eastern and south-eastern Australia and Tasmania.

Birds roosting at night probably make up a large part of its prey, but it is able also to kill small wallabies. Together with the smaller native-cats and phascogales, this carnivore is undoubtedly responsible for many of the losses suffered by nesting birds — but this of course has always been a part of the balance of nature in the Australian bush.

There are many tales of the boldness of the Tiger Cat. The presence of one of these carnivores in the district is usually first noticed when there are nocturnal killings of roosting fowls. If trapped or cornered the Tiger Cat, with its widely opening jaws and slashing canine teeth, is not to be despatched easily. One was reported to have kept two terriers at bay from its vantage point in a hollow; another killed a large tomcat in a savage fight.

The tree-climbing skill of the Tiger Cat is largely due to the adaptations of the hind feet, which have sharp claws and serrated rubbery pads on the soles.

Four to six young are born about May, and carried in a rudimentary pouch, a crescent-shaped flap of skin enclosing the front and sides of the mammary area.

The large Tiger Cat is in greater danger of extinction because of the boldness of its raids on fowl pens, returning persistently on subsequent nights and showing such lack of fear, when hunting and feeding, that trapping or shooting becomes almost inevitable.

But in spite of the bloodthirsty raids upon roosting hens by an occasional Tiger Cat, the species on the whole is beneficial, helping to keep rabbits, rats, mice and other pests under control.

Much smaller than the Tiger Cat are the native-cats, which differ also in having unspotted tails, sharply pointed faces, and relatively larger ears. There are three species.

Tiger Cat

The Eastern Native-cat or Quoll inhabits forests along the coasts and mountain ranges of eastern and south-eastern Australia; the Western Native-cat or Chudich was once very widespread in western New South Wales, southern Queensland and north-western Victoria through Central Australia to coastal Western Australia; and the Little Northern Native-cat occurs in tropical north-coastal Australia.

The Quoll, with a head-and-body length of about 45 cm and a tail of 30 cm, is more a terrestrial predator than the Tiger Cat; this shows in the structure of its feet, which lack the strongly ridged sole-pads of the larger predator.

During the day the Quoll sleeps in a nest under rocks or in a hollow log, becoming active towards evening. It stalks its prey carefully and patiently before springing to kill with a quick bite across the back of the head.

Compared with the domesticated placental cat, the head of the native-cat is more elongated, giving a sharp-faced appearance, but the eyes and ears show similar adaptations for nocturnal hunting. The ears are erect and can be directed both forwards and sideways, and the hearing is very acute. The eyes are large for the size of the head, with wide pupils to gather in the weakest light.

The fur is soft and thick, but usually shorter than that of the placental cat. It may be coloured black, brown or olive-grey, with prominent white spots on body and head, but not on the tail.

Native-cats have a shallow pouch in which there are six, sometimes eight, teats. As many as twenty-four young are born, but only the first six or eight that attach to a teat can survive. These remain attached for two months. Their eyes open at eleven weeks, they are weaned at four months, and become independent at around five months of age.

The insectivorous marsupial-mice are divided into two main groups — the broad-footed and the narrow-footed species. The former, as the name indicates, have short, broad feet. Their tails are short-haired and tapering, never crested or brush-tipped. As insects form the major part of their diet they have a full set of pointed incisor teeth typical of the carnivore. Included in the broad-footed group are the Dibbler, the Yellow-footed Marsupial-mouse, and three species of *Planigale*.

The Planigales, or flat-headed marsupial-mice, are the smallest known species of marsupial, and among the smallest mammals in the world. So small in fact that they are close to the lower limits of size postulated to exist for warm-blooded animals.

Small mammals need a lot of food because they loose body heat (derived from food) very quickly, as their surface area is very large in proportion to their body weight. Very small animals have a larger surface to body-volume area than larger animals. They find it difficult to obtain enough energy to keep their body temperature constant, and have very small reserves of energy. Thus the tiny *Planigale* usually eats a quantity of insect food at least equal to the volume of its own body each day, and cannot go long without eating.

One species, the Kimberley Planigale, is very much smaller than an ordinary mouse, having a head-and-body length of 38 to 50 mm. It is scarcely bigger than a large grasshopper, so quite a prolonged battle ensues when it attacks one of these strong insects.

A peculiar characteristic is the flattened head, which is little more than 10 mm from base to crown.

Very little is known of the habits of the four species of flat-headed marsupial-mice. A female with seven tiny young, kept in captivity, made a small saucer-shaped nest of dried grass in the concave space between a piece of wood and its loose bark. She ate more than her own body weight in insect food each day, particularly grasshoppers, and also ate small dead birds.

So little research has been done that almost nothing is known of the biology of these pigmy marsupial-mice in the wild. They were once thought to be extremely rare, but in the case of at least one species, the Kimberley Planigale, the scarcity of specimens is probably due more to the ruggedness and inaccessibility of the habitat, rather than lack of numbers.

The narrow-footed group of marsupial-mice contains many tiny carnivores; most species are about the same size or slightly larger than the ordinary house mouse. They can be separated from the preceding genera of broad-footed marsupial-mice by the appearance of the hind feet, which are long and very narrow. The ears are much larger, and the pouch better developed. They are active hunters, attacking and devouring insects and other small creatures at night.

Especially interesting and attractive little animals of this group are the two species of jerboa marsupial-mice. These long-legged, long-tailed, little, insect-eating marsupials (not related to the rodent hopping-mice) were once thought to travel like little kangaroos, leaping. But studies using high-speed photography showed that their movement is a fast succession of leap-frog actions, with fore-feet touching the ground fractionally ahead of the hind feet, in a galloping action.

Below:

The Chudich or Western Native-cat, *Dasyurus geoffroii*, is much smaller than the Tiger Cat; being only about thirty inches in length included the unspotted tail. With claws and teeth it rips apart a parrot which it has caught by surprise at night. Formerly found through most of inland and western parts of the continent, it now seems to exist only in the forests of south-western Australia. Two similar native-cat species inhabit the forests of eastern and tropical northern Australia; unlike the Tiger Cat, these native-cats have spotted tails.

Unique to Tasmania, the Devil in the early days of settlement acquired a reputation of being a sheep-killer. The Tasmanian Devil is probably the most ferocious looking, and sounding, of all Australian marsupials. Although just over a metre in length, the Devil is extremely heavily and powerfully built. It has become more common in Tasmania in recent years.

4 Relics of the Past: Platypus and Spiny Anteater

From the prehistoric reptiles of the distant past has evolved a large part of the animal life as we know it today. Those early reptiles gave rise on one hand to the modern reptiles, and, on the other, branched out into a separate new line of evolution — the first primitive mammals.

Those earliest mammals have in turn long since vanished; from their fossilised remains in the rocks we can see that they were quite unlike the mammals of today, for they retained very many reptilian characteristics.

The development of mammals then branched out along three different paths. Each group eventually became quite successful in its own way, diversifying into many highly specialised species. But some were built on a basically more successful, more efficient plan than others, and gradually became dominant.

These three almost parallel lines of evolution, which mammals have followed since the time of the first reptile-like mammals, have culminated in three orders of mammals, all represented by living species on earth today — the monotremes, the marsupials, and the placental or eutherian mammals.

The monotremes, which retained many of the reptilian characteristics of the earliest mammals, appear to have been the least successful, for they were displaced almost entirely by the marsupial and placental mammals. The marsupials, too, vanished from most parts of the earth as the placental mammals became more and more efficient, breeding faster, surviving more successfully the adversities of their environment. Only in isolated Australia and New Guinea, drifting far from the other placental-dominated continents, did the marsupials remain the major part of the mammal fauna; only on these islands did any of the primitive monotremes survive at all.

Although all three orders of mammals had their beginning in the time of the early mammals, the monotremes remained much more like their ancestors. This group contains the most primitive mammals living today, still retaining numerous reptile-like features. They represent a poorly developed branch of the mammalian evolutionary tree.

The monotremes combine certain features of reptiles, marsupials and placentals. From the reptile skeleton they retain the pectoral or shoulder girdle bones, and reptile-like skull; they have a reptilian eye structure, a single combined genital and excretory opening (the meaning of the word monotreme), and they lay soft-shelled eggs like those of reptiles.

Characteristics shared by both monotremes and marsupials are the epipubic bones in the pelvic girdle, and the lack of any connections within the brain between the two cerebral hemispheres.

Unlike reptiles, but in common with marsupial and placental mammals, they have milk-producing mammary glands, a body covering of fur, a more efficient four-chambered heart, and a single lower jaw structure.

In the past it was assumed that another indicator of the close affinity between reptiles and monotremes was the poison spur on the hind limbs of the males. However there is no reason to regard the poison spur as showing any physiological affinity with the venomous reptiles.

Reptiles are unable to control their body temperatures except by behavioural tactics such as basking in the sun, but mammals can maintain a steady warm body temperature by purely internal means. The monotremes have not quite achieved a control comparable to other more advanced mammals; their temperature averages somewhat lower at 31.1° C and fluctuates between 22.2° and 35.5° C.

One of the world's few living species of monotreme, the Spiny Anteater shares with the Platypus the distinction of being a primitive link between the reptiles and the mammals. The anteater and the Platypus are the only mammals to lay eggs, and have reptile-like features of reproductive system, skeleton and body temperature regulation. Although the Spiny Anteater is very widespread, from rainforest to semi-desert, it is an inconspicuous creature. When disturbed its immediate reaction is to dig vertically downwards, and once half buried, with powerful claws gripping among the roots, it is very hard to dislodge.

Left:

If taken by surprise, or on a surface too hard for digging, the Spiny Anteater curls into a tight ball, protecting its face and undersurface where there are no spines.

Below:

Probably the strangest of all Australian animals, the Platypus at first appears to be a peculiar mixture of bird and mammal. But such unusual features as the duck bill and webbed feet are adaptations to the amphibious way of life, while egg-laying rather than live birth is a characteristic retained from its not-so-distant reptile ancestors. In the water its webbed feet expand into broad paddles, though the hind feet usually trail behind giving, together with the tail, stability and directional control. When it dives the ears and eyes are closed; the prey is found and identified by touch, using the sensitive bill.

Probably the greatest single gain by the monotremes when compared with the reptiles is their much larger brain — as large as that of the more primitive of the marsupial and placental mammals.

The world's only living monotremes, the Platypus and the echidnas or spiny anteaters, are endemic to the Australian region. The Platypus is confined to eastern Australia, and the spiny anteaters to Australia and New Guinea. Of these anteaters there are five species in two genera. One species, *Tachyglossus acculeatus*, is found throughout both Australia and New Guinea, and the other four species in New Guinea.

The Platypus is often described as 'duck-billed', but this is a very superficial comparison. The beak or bill is not horny like that of birds, but is instead a framework of bone covered with soft, moist skin naked of any fur. The bill increases in width towards its rounded tip, where the pair of nostril apertures are situated. The rest of the head is furred, and merges into the body without any clearly defined neck.

The Platypus has no external ears, and the eyes are small; both the ear apertures and the eyes are set into a deep groove in the skin, the edges of which close together when it is swimming underwater. This probably renders the Platypus both deaf and blind when beneath the surface. But it is in that environment that the marvellous bill takes over, guiding the Platypus along the waterways and creekbeds, around boulders, under and over logs and among reeds, finding and identifying the prey. The skin of the bill contains a mass of very sensitive nerve endings. Although not used underwater, the sight and hearing are reputed to be keen.

While generally thought of as aquatic animals, Platypuses actually spend only about two hours out of every twenty-four in the water, when they feed.

The Platypus emerges from its burrow in the early morning and late afternoon. In the water its webbed front feet expand into broad paddles. The hind feet, also webbed, trail behind, and together with the tail, give stability and directional control. The Platypus surfaces often to breathe, and to chew its prey. It appears that the prey is found and identified solely by touch, then snapped up and stored in cheek pouches, until brought to the surface.

Platypuses when feeding stay submerged for periods of one to five minutes. When floating on the surface they can look so much like a piece of floating debris that often they are not noticed. The Platypus excavates a burrow which is usually 4.5 to 9 m, but sometimes as much as 27 m in length; the entrance is well above water level. At intervals it is plugged with soil, perhaps to deter predators, and possibly also to maintain a warm humid atmosphere for incubation.

In the subterranean chamber a nest of grass and gum leaves is constructed, and two soft-shelled eggs laid. The Platypus, like the Echidna, has no nipples, the milk being sucked up by the young as it exudes from the pore-like ducts of the mammary glands.

Platypus

The Spiny Anteater *Tachyglossus aculeatus*, although belonging to the most primitive group of animals in the world, has remained one of the most successful of all Australian furred animals. It is extremely widely distributed throughout Australia and Tasmania, being common over a wide range of habitats from semi-desert to rainforest. The Spiny Anteater's long tube-like muzzle, like the 'beak' of the Platypus, is an adaptation to a way of feeding. The tiny mouth opens just wide enough to allow the rapid probing action of the long tongue in picking up ants and termites.

Confident of the protection given by its coat of spines, the anteater is out during day-time, and most active in early afternoon.

If it is disturbed, the anteater's immediate reaction is to burrow straight downwards, and once half buried, with the heavy claws of its immensely strong limbs hooked under roots or rocks, it is almost impossible to dislodge. But if the anteater does happen to be taken by surprise on a hard surface where it cannot burrow or grip, it curls into a tight ball, protecting its face and underface where there are no spines.

The generic name of the Spiny Anteater, *Tachyglossus*, means quick-tongued. The small mouth opens below the tip of the beak-like muzzle, and the tongue can extend about 30 cm. It is sticky, and flicks rapidly in and out during the meal of ants.

Most of the Anteater's head and undersurface is covered with hair, but small spines cover the back of the head, and there are very long sharp spines all over the upper surface and sides of the body.

The female Spiny Anteater lays a single egg, which is put into a temporary pouch on the undersurface of her body. The anteater apparently lies on its back and curves its body so that the egg, perhaps with some guidance from the snout, rolls into the pouch.

After seven to ten days in the pouch the egg hatches. The tiny hairless baby lives in the pouch for six to eight weeks, until its new spines become uncomfortable for the mother, who removes it to a safe hiding place until it is able to fend for itself.

5 Land of Colourful Birds

Australia has about 700 species of birds, most of which are found in no other part of the world. More than 500 of these are land birds and breed within this continent. Some bird families are confined entirely to the region, and other groups have their greatest radiation of species within Australia. Many families that are widespread in other parts of the world have, in Australia, some magnificently coloured species which are unique to this island continent.

Among the families in which Australian species are dominant are the parrots (about sixty species), honeyeaters (about seventy), and finches (twenty). Other groups of colourful birds unique to Australia and nearby islands are the fairy wrens, many flycatchers including robins both red-breasted and yellow, and the vividly coloured chats.

Australia's many blue wrens, or fairy wrens, form a particularly colourful group. Although colour varies greatly from species to species, with some having no blue at all in their plumage, their very long tails, held vertically most of the time, make these birds very easy to recognise.

These fairy wrens, genus *Malurus*, are placed in the Old World warbler family, *Sylviidae*, sub-family *Malurinae*. The fourteen species occupy a wide range of habitats covering most of the continent. Best known is the ordinary Blue Wren of coastal south-eastern Australia. In pairs or small parties it inhabits bushland undergrowth, and is often a visitor or resident in gardens.

Fairy Wrens will be seen in dry and semi-desert inland areas — the Blue-and-white Wren, the Black-backed Wren, the Purple-backed Wren and the Variegated Wren. Others favour the tropical northern and north-eastern coastal strip — the Lilac-crowned Wren, the Red-backed Wren and the Lovely Wren. Some are isolated in distant corners of the continent: the Splendid Wren and the Red-winged Wren of south-western Australia. Another, the Black-and-white Wren, occurs only on two islands off the north-west coast.

The Australian robins belong to the Old World Flycatcher family, *Muscicapidae*, but unlike the typical flycatchers which form part of that family, the robins of Australia (sub-family *Muscicapinae*) do not usually capture insects in the air. They sit, motionless but for an occasional flick of the wings, on a perch, or clinging to the side of a tree trunk, then drop to the ground to take their prey.

There are five species with red or pink breasts, some six species with extensive areas of bright yellow, and a number of less colourful birds including the White-breasted Robin, the Dusky Robin, the Hooded Robin and others.

Characteristics of these robins of the Old World Flycatcher family include relatively weak legs and feet (being used only for perching), ten primaries in the wings, and pronounced bristles at the base of the beak. Their nests are open, cup-shaped structures, which in the case of the robins are beautifully camouflaged with bark, mosses and lichen, bound and matted with spider webs.

Fantails fit the flycatcher image much better than the sedate robins, being active birds, always restless when perched, constantly fanning out their long tails, twisting this way and that, then suddenly fluttering up among the foliage, twisting and turning in complicated aerial acrobatics in pursuit of a fly or other small insect. In Australia there are two species: the Grey Fantail, and the beautifully coloured Rufous Fantail which inhabits the forests of coastal eastern Australia.

Sometimes grouped with flycatchers are the whistlers, somewhat larger and slower birds than the robins and fantails. The Golden Whistler is the most showy of the genus, the male being coloured crisp black and

The Rufous Fantail, a small bird of the flycatcher family, is usually seen in rainforests and fern gullies around Australia's eastern and northern coasts. Constantly on the move, fanning out its long tail and twisting restlessly from one side to another, the Fantail brings a patch of bright colour and movement to the green gloom of the jungle.

Crested Pigeon

white with breast and underparts a bright golden yellow. His powerful calls are among the commonest of bird sounds in coastal forests around Australia in spring months.

Australia's many colourful parrots, cockatoos and lorikeets together make up one of the best-known and most distinctive of bird groups. The Order Psittaciformes contains three families which share the basic parrot features, which include a short rounded bill with hinged and movable upper mandible, nostrils set in a fleshy cere at the base of the bill, and a thick prehensile tongue.

Few birds have colours to surpass the gaudy rainbow colours that adorn some of the Australian parrots. There are the numerous species of rosellas, many, like the Western Rosella, with plumage of crimson, yellow, blue, green and white.

Cockatoos, the largest of Australia's parrot-like birds, have a conspicuous crest, often brightly coloured. Another family, the lorikeets, has many brightly plumaged members, of which the Rainbow Lorikeet is one well-known example. The lorikeets are brush-tongued parrots which feed largely upon the nectar of flowering trees and shrubs. It is among the many parrots, perhaps more than any other bird group, that the plumage of Australian birds reaches a peak of brilliant colour, making this a truly spectacular segment of the birdlife.

Spotted Harrier, a beautifully coloured bird of prey

Right:

On his way to his nest a male Red-winged Wren alights on gumnuts among the ferns under a forest thicket. The chestnut coloured shoulder patches which give him his name are partly hidden by his raised wings. Red-winged Wrens are unique to south-western Australia, where they favour dense vegetation near forest streams.

Far Right:

A male Mistletoe Bird arrives with a beakful of sticky mistletoe berries. The nest, which looks and feels like a woollen baby's bootee, is made of soft plant down, empty spider egg sacs, matted together with cobwebs until it has the strength, softness, texture and appearance of knitted wool. It is hung from a thin leafy twig in the topmost foliage of a tree or shrub. The sticky sweet fruits of the mistletoe are the main food of these birds.

Right:

A male Splendid Wren, long tail streaming behind, flies through the forest undergrowth. This species is unique to south-western Australia.

Below:

A male Red-backed Wren throws his wings up for take off. These birds are an incredible sight during their courtship display flights, when the crimson feathers are fluffed out to spread right across back, shoulders, and sides, turning them into living balls of fire as they fly slowly and conspicuously well above the scrub.

Far Left:

A richly plumaged male Golden Whistler drops down to land beside his mate at the nest. Golden Whistlers are found in Australian coastal forests, where their loud ringing calls are among the most common of springtime sounds.

Left:

Black-faced Flycatchers are birds of the rainforests of coastal eastern Australia. Their deep cup-shaped nest is covered with green moss so that, even from a short distance, it can hardly be distinguished from the countless tangles and clumps of the same moss which hang from the branches of the rainforest trees.

Below:

Crimson Chats are one of the brightest coloured of the small desert and arid-land bird species in Australia. Their nest is well hidden in a violet-flowered mulla-mulla bush, and although only about 10 cm above the ground, is almost invisible. These dense bushes are the chat's favourite nesting site in north-western Australia.

6 Bird Wonders of Australia

The Australian region has many birds which have no equal in any other part of the world. During the long period of geographical isolation they have each followed a particular path of evolution, becoming ever more and more specialised. A few are relic forms of birdlife, representing a very old element within the Australian bird population. Some are of particular interest for their unique behaviour patterns, especially in courtship. Others have developed remarkable methods of incubating their eggs. The bowerbirds and the incubator birds in particular have fascinated scientists ever since their discovery; only in recent years have the details been revealed.

Bowerbirds are unique among the birds of the world in their skill as architects, the use of tools, and use of colour in decorating their structures.

There are nineteen species, six being found only in Australia, eleven being restricted to New Guinea, and two common to both Australia and New Guinea. Most bowerbirds inhabit regions of mountain ranges and dense jungles, where observations are difficult and photographs rarely obtainable; but in Australia several species live in semi-desert regions.

Bowerbirds vary considerably in size and shape; some are extremely brightly plumaged, others are very plain. Their most constant characteristic in common is the bower, rather than any particular feature of body or plumage.

This bower is a specially prepared dancing stage or platform on the ground, or a decorated perch above the jungle floor. Some are simply clearings on the ground, while others are elaborate structures with walls or towers of sticks, decorated with a variety of natural treasures, such as snail shells, colourful feathers, bleached bones, seeds, brightly coloured native fruits, flowers and mosses. In Australia, where bowerbirds sometimes occur not far from human habitations, their bowers commonly contain a variety of man-made ornaments carefully selected to conform to the colour scheme of the bower — coloured glass, small colourful plastic items, shiny metal and the like.

The crowning achievement of bowerbirds is the mixing of natural pigments and the painting of the walls of the bowers. Two of these species use tools, crudely fashioned brushes, to assist in applying the colour, and thereby join the ranks of the few animals that deliberately use a tool to achieve a desired end.

It was once thought that bowers, with their elaborate architecture and decorations, were merely playgrounds in which the males amused themselves, apparently out of sheer exuberance.

But each bower is the focal point of a male's territory. He begins to build or repair it each year in the season when he comes to a sexually active state, and maintains the structure only as long as this condition continues — a period of three to five months.

The females, however, may remain sexually inactive during much of this period, until the conditions of the environment (rainfall, heat, abundance of food and so on) reach the levels necessary to trigger the breeding cycle. When this happens, the female becomes sexually active, accepting the male in his bower.

According to this theory, the females are triggered not by the behaviour of the male, or by the bower, but by environmental conditions most favourable for a successful nesting.

The ornithologist E. Thomas Gilliard, who studied bowerbirds extensively in New Guinea, has put forward a 'transferral theory' to explain the building of the bowers. He noticed a correlation between the elaboration of the bower, and the development of special courtship plumage: the best bower-builders tend to have the dullest plumage.

According to this theory, bowerbirds with brightly coloured, long trailing plumes or large crests, when displaying on the ground are very vulnerable to predators. Those that build an elaborate bower, decorated

A male Satin Bowerbird at the bower which he has constructed, decorated and painted. The bower is the focal point in his territory. It is not a nest, but a stage where he displays to the female, and where mating occurs if she is sufficiently impressed by the bower and his courtship display to be enticed between its walls.

with colourful objects, are in effect transferring the role of optical stimulation of the female from themselves to the bower. The bright plumage therefore becomes redundant, and gradually lost through natural selection, because those individuals with smaller crests and duller colours tend to live longer and breed more often.

It is thought that the Green Catbird of Australia and New Guinea is the most primitive of the bowerbirds. It has brilliant green plumage patterned with white and black markings, and does not build any bower. The male and female, which have similar plumages, both attend the nest.

More advanced in the development of a bower is the Tooth-billed Bowerbird of north-east Queensland's mountainous jungles. The male of this dull olive coloured species cuts off leaves with his serrated-edged bill, and spreads them across the ground to create a 'stage', usually about 1.5 m in diameter. The tooth-bill is an accomplished ventriloquist, and sings loudly much of the day from a favourite perch over his display area.

The smallest of the bowerbirds is an Australian species, the Golden Bowerbird, yet it builds an enormous structure which may be as high as 2.7 m. It is re-used each year, so the biggest examples of such bowers would be the result of many years of building and additions, but nevertheless a remarkable piece of architecture for so small a bird. The collection of sticks is usually piled around several slender saplings which stabilise the whole structure, while a horizontal branch or stick between the two towers serves as the display perch. The bower is decorated with yellowish lichens and golden orchid flowers.

The male Golden Bowerbird, like other males of the group, takes no part in nest building or rearing of young.

The most highly developed of the bowerbirds are probably the avenue-builders. Within Australia these include the Satin Bowerbird, the Regent, the Spotted, the Fawn-breasted and the Great Bowerbird.

The male constructs his bower by first building up a meshed horizontal platform of sticks, then inserting sticks vertically to build two parallel walls about 40 cm in height. These walls are just wide enough for the birds to move between. Each species decorates its bower with colourful objects — the iridescent dark-blue Satin Bowerbird collects predominantly blue objects, but also some yellowish flowers.

The Satin Bowerbird, like several other species, has developed the fascinating practice of painting the inner walls of the bower. It chews up charcoal or decayed wood and, using a small wad of bark held in the tip of the bill, wipes the sticky brownish mess up and down the veritical sticks of the wall. This is one of the rare instances of a bird intentionally using a tool.

It is not known what advantage is gained by building bowers, but it is thought that these birds are polygamous, and that a pattern of behaviour has evolved which encourages the females to visit the males, rather than the reverse.

Regent Bowerbird

The remarkable incubator birds, or megapodes, bury their eggs in the primeval manner of the reptiles. They possibly had their origins in New Guinea, and spread northwards among the islands, and southwards into Australia where three species occur.

The megapodes are the only known birds to use heat other than that of the parents' bodies for the incubation of their eggs. Three different natural sources of heat are utilised — the warmth generated by rotting vegetation, heat accumulated from the sun, and (not in Australia but to the north on the Celebes Islands) the warmth given off by volcanic activity deep underground.

Three species of megapode occur in Australia: the Scrub Fowl, the Brush Turkey, and the Mallee Fowl. The first two of these are 'conventional' incubator-birds, living in warm humid jungles of north-eastern Australia where conditions are ideal for compost-heap nesting.

The Brush Turkey is a well-known bird which has become quite tame and easily seen in many areas of jungle habitat in the national parks. This large bird is black with a naked head and colourful neck wattles of red and yellow.

The incubator-mound is a huge conical pile, often 3 m high, of leaves and twigs, which the birds have raked up with their powerful feet from a large area of the forest floor. In this mound, where heat generated by the compost reaches about 35° to 36° C, seven to twelve eggs are laid at a depth of about a metre. When, after a long incubation period, the eggs hatch, the young turkeys must dig their way to the surface. Their feathers are free of their waxy sheathings, and they are immediately able to run and flutter into the air. They are quite independent of their parents and each other.

The most amazing of all incubator birds, however, must be the Mallee Fowl. This large bird inhabits mallee eucalypt and similar dry scrub and open forest country across southern Australia. In this environment

Brush Turkey

the rainfall is much less than in the tropical rainforests, and moreover occurs during the cool winter months, when heat-generating fermentation is minimal. For the Mallee Fowl it is not sufficient simply to cover the eggs and forget them. The tending of the buried eggs, to maintain an almost constant temperature, is one of the most incredible of bird achievements.

The pattern of behaviour involved in the hatching of Mallee Fowl eggs is fantastically complex. The tending of the mound occupies the pair of birds eleven months of the year. Beginning in April or May with the first showers that herald the approach of winter's general rains, the pair of Mallee Fowls dig out a wide hole, and scratch into it a compost heap of wet leaves and other forest debris. By mid winter this pile has attained substantial proportions, and has begun to rot. It is then covered with sand to a depth of about a metre to seal in the heat.

The female begins to lay in the early spring, the male opening a hole for her in the top of the mound, which by now may measure 4.5 to 6 m in diameter, and 1.2 to 1.8 m in height. Eggs are laid at intervals of a few days, and laying continues for the next five or six months.

Meanwhile the male attends the mound almost constantly. His task is to maintain the temperature of the egg chamber at 32° to 35° C. Somehow he is able to judge the temperature by his beak, which he pushes into the mound as though it were a thermometer.

If the temperature is not correct, he follows a pattern of actions appropriate to the circumstances.

In early spring months, when the heat from the moist compost under the eggs is greatest, it will tend to become too hot during the middle of the day. The male Mallee Fowl must be at work before daybreak, opening out the mound and spreading several cubic metres of sand to cool in the chilly morning air; before the sun breaks through the trees he must pile all this sand back on top of the mound. However, if the weather is cold, he may leave the mound closed up, or possibly add more insulating sand to keep in the heat.

In mid-summer there is still some heat being derived from fermentation, but also a great deal of heat from the sun, so the mound is kept covered with a thick layer of sand to insulate against the sun heat. Once every day or two, when the heat builds up above the optimum level the mound is opened before dawn, and the sand spread out to cool. The cooled sand is piled on thickly soon after sunrise, to insulate against the heat of the sun throughout the day.

A third technique is adopted in autumn, when there is no heat being derived from the now dry compost, and little from the sun. Each day the mound is opened out around 10 or 11 a m., leaving only 5 to 7 cm of sand over the eggs. Throughout the hottest part of the day the exposed interior of the mound soaks up the sun's heat. The male Mallee Fowl returns the sand in shallow layers, allowing time for each layer to heat up, until by mid afternoon the mound is fully built up, and its internal temperature is near 34° C.

The chicks hatch beneath about a metre of sand, and dig upwards to the surface within two or three hours. They can run and flutter, and fly quite well by the end of their first day. They are solitary, and completely independent of their parents, fending for themselves in every way from the time they hatch.

Lyrebird singing and displaying

Left:

The Great Bowerbird carries a snail shell through the bower and adds it to the pile of greyish objects. Just as the Satin Bowerbird selects blue objects, making his bower an extension of his own blue colour, so the grey-brown Great Bowerbird decorates his bower with white, grey and brownish objects.

Below:

With the female entering the far end of the bower he has tossed aside the ball of silver paper and holds a metal ring; his posture and position are the same, and he slowly twists his head from side to side to draw attention to the display object. Mating occurs when the female has been enticed into the bower.

The Golden Bowerbird is the smallest of the bowerbirds of Australia and New Guinea, yet builds one of the biggest bowers, a massive pile of sticks stacked up around two slender saplings 30 to 60 cm apart. Bridging between the towers is the display perch. The sides of the perch are thickly padded with golden moss while yellow orchids and other flowers are threaded among the sticks of the towers. This is another instance of the bowerbird choosing decorations of the same colour as its own plumage, making the bower an elaborate extension of its own colour scheme. The Golden Bowerbird is unique to Australia, inhabiting mountainous parts of north-eastern Queensland.

The most amazing of the incubator birds is Australia's Mallee Fowl. This large bird inhabits mallee eucalypt and similar dry bushland across southern Australia. The Mallee Fowl's tending of its eggs, buried in the massive incubator mound, is one of the most incredible of bird achievements. The pattern of behaviour required to keep the eggs at an almost constant warm temperature is incredibly complex for bird behaviour; the tending of the mound keeps them busy eleven months of the year. Most of the work is done by the male, who must keep the temperature of the egg chamber at 32° to 35° C; using his beak as a thermometer he checks the mound and regulates the heat gain or loss.

Right:

Perched in a bush not far from the entrance of his nest tunnel, a male Spotted Pardalote spreads wings and tail in a threat display intended as a warning to a rival male who has entered his territory. The Pardalote turns his back towards the intruder (but turns his head to watch the other bird) while spreading wings and tail to show the colours and spotted pattern of his species. For a few seconds he holds this pose, while swaying slowly from side to side, then, if the other bird has not retreated, suddenly dives upon him, beak clicking in a bold exhibition of ferocity, chasing him away.

Below:

The most archaic of the birds of Australia are the huge flightless Cassowaries and Emus. These birds of primitive origin may have begun their evolution towards flightlessness and giant size in an era of the distant past when mammals offered less competition than they do today. The Emu stands 1.5 to 1.8 m tall and weighs as much as 150 kg; it can manage speeds near 65 km per hour. Once occurring almost throughout Australia, Emus are now absent from most settled coastal areas, but fortunately remain common on the vast inland plains.

Although the Spotted Pardalote is a bird of the tree tops, where it lives upon tiny insects around flowers and on the foliage, it chooses to nest in a damp dark underground chamber. The male and female together scratch a narrow tunnel into a creek bank and build in it a globular nest of grass and bark. The male shares in the incubation of the eggs, and when the young hatch, helps his mate maintain the supply of insects which they collect from the trees of the forest above.

7 Unique Partnership

There exists between the plants and the animals of the Australian bushland a unique partnership. Many birds, mammals and insects live upon the nectar and pollen of native flowering trees, shrubs and small plants, while these plants which have provided the lures of abundant nectar and pollen are entirely dependent upon the comings and goings of the bush creatures to carry their pollen from flower to flower for cross-pollination.

The flowers on the one hand have evolved ingenious structures to display and advertise their 'bait', and to ensure efficient transfer of the pollen on the body of the pollinator. Some of these devices, particularly among orchids, are amazing almost beyond belief. The animal world, on the other hand, presented with a rich source of food, has in a multitude of forms, evolved creatures that specialise in nectar-eating.

Many wildflowers, and orchids in particular, have chosen, or evolved with, a small group, or even a single species of pollinator. Very often the nectar is inaccessible or unattractive to any but the one species; in many instances that creature will visit this flower in preference to all others, so that there exists these closed chains of plant-animal cooperation, of great benefit to both. There is the assurance that the pollen will be carried from flower to flower within the one species of plant, avoiding waste of pollen on other flowers where it is useless. The pollinating animal has the advantage of an exclusive nectar supply, which his species alone can find or reach.

It is fascinating to examine the lengths to which the flowers on the one hand have gone to achieve this end, and on the other, the adaptations wrought by millions of years of evolotion upon the animals which are a part of the unique partnership.

Some plants are pollinated by the haphazard scattering of their pollen in the wind, a dispersal method that is wasteful and inefficient compared with pollination by animal agencies. Most of the larger-flowered species have evolved their present shape, colour, size and texture under the influence of the bush creatures that serve to carry their pollen. Consequently there are vast differences between flowers evolved for bird, mammal or insect visitors.

Flowers which would catch our attention from a distance would just as likely be an attraction to honey-eating birds, whose sight is similar, but may not necessarily be the most conspicuous to insect eyes.

Reds and yellows stand out in striking colour contrast against the grey-greens of bushland foliage, catching the attention of bird and man alike. But to insect eyes a flower that we would describe as a dull blue-violet could well be as conspicuous as a scarlet bottlebrush would be to us. Colour among wildflowers is not a haphazard or accidental thing, but serves the vital purpose of attracting the desired pollinator.

Any flower which is coloured red to yellow, has a large strong rigid spike or head, and has its pollen dispensed and received 25 mm or more from the source of nectar, may be suspected as being bird pollinated. Many of the large, spectacular Western Australian wildflowers could not possibly be pollinated by insects. Some species appear to effectively shut out all insects from their nectar, as they require the strong pointed beak of a honeyeater to spread open a constricted opening or push through a barrier of stiff hairs. Other flowers simply discourage insects by omitting all the facilities offered by the insect flowers. Other flowers again allow insects, birds and all others a free go at the nectar, even though it may be only the birds that have the correct size and shape to serve as pollen-carrying agents — insects coming for nectar are themselves an attraction for honeyeating birds, which are also insectivorous.

It is thought that birds may be even more red-conscious than man, when it comes to seeing this colour against the grey-greens of the bush landscape. Some

This Yellow-faced Honeyeater seeking nectar at a Christmas Bell will carry pollen from plant to plant. The bird must thrust its head deep into the flower, brushing the style and pollen-covered anthers; the plant's stiff stem provides a very convenient perch.

birds see as though through a red filter, having microscopic red or yellowish oil droplets scattered through the retina of their eyes, absorbing some green light, and thereby darkening the green areas of the image.

Colour among insect-pollinated flowers too is functional. Although it is hard to say exactly how an insect views the colour pattern of its world, we know that insects have very definite preferences, and that insect flowers are very often blue and violet and white. Their choice of the 'cool' end of the spectrum contrasts with birds' preferences for the warm colours. There are exceptions, and some overlap; yellow is common in both insect and bird floral designs.

White, blue and even green flowers, set among green foliage, attract insects, even though any influence of scent may be experimentally excluded. There are fundamental differences between insect colour vision and that of the higher animals: insects recognise ultraviolet light as a visible colour, but seem usually to be blind to red. Their colour spectrum is substantially different, red appearing black, and ultra violet being a visible colour. Thus red flowers would not be the the bright objects they are to us.

But many flowers looking red, or any other colour, to us, may be reflecting a lot of ultra violet which we do not see, so flowers of any colour can and probably do look different, as far as relative brightnesses are concerned, to insect eyes. This affects not only the way a flower stands out against the background, but also the contrast of dots, lines and other markings.

Flowers pollinated by insects usually provide a convenient landing petal which is often spotted and streaked with markings of various colours showing the way to the nectar. Karl von Frisch, the pioneer researcher on the subject of insect communication, studied the colours of petals and of their nectar-guide markings, and found that the commonest combinations to be those most easily distinguishable by bees — yellow and purple, orange and blue, yellow and violet. These nectar guides serve to direct the insect visitor from the landing platform to the nectar, and the marks are most elaborately developed when the nectar is deeply hidden.

If the birds have influenced the shaping of the large wildflowers, then insects have equally played a part in shaping the very small flowers, and it is the orchid family that has developed the most complex and efficient pollination systems. But of all the cunning devices used to attract insects, one of the most amazing must be the slipper or tongue orchid, of which there are a number of rather similar species around Australia.

The Slipper orchids, genus *Cryptostylis*, appears to have only one species of pollinator, an ichneumon wasp, and it invites only the male wasp. This flower has evolved, as a lure, an attraction that duplicates the attraction of the female wasp, so that the male instinctively responds and comes to mate with the imitator.

Spinebill perched on Kangaroo-paw

While the male wasp clings motionless beneath the red-veined slipper-shaped labellum of the flower, the tip of his abdomen is in contact with the orchid's pollen sac, or pollina. This sticky golden mass of pollen becomes firmly glued to him; when he eventually flies off he must carry it wherever he goes. Sooner or later he will come near another orchid of the same species, succumb to its charms, and in so doing, deliver the pollen, effecting the cross-pollination of the orchid.

The birds, mammals and insects which specialise in feeding on nectar have evolved special adaptations to match those of their age-long partners, the flowers. The principal nectar-eating birds are the honeyeaters, and the lorikeets, unique to this southwest Pacific region.

Australian honeyeaters belong to the family Meliphagidae, containing about 160 species of small to medium-sized nectar-eating birds. Their beaks are always fairly slender, often long and usually downcurved. The tongues are extensile, and brushlike.

The wildflowers provide honeyeaters directly with only one of their two essential food requirements, nectar, usually in the form of sucrose. But indirectly the flowers also provide much of the protein, consisting of insects attracted to the nectar, small spiders and the like, taken from the flowers or caught in the air around the flowers.

Tawny-crowned Honeyeater feeding

The most obvious modification of honeyeaters for the liquid nectar diet is the tongue. This is curved up along its sides to form a long tube which can be thrust out 6 to 12 mm, and may fill automatically when thrust into nectar because the act of protrusion forms a vacuum between the base of the tongue and the palate.

The honeyeater tongue may be fringed or brushlike. Its tip seems to oscillate rapidly in use, wiping, lifting and pumping the nectar.

Birds have excellent sight and good colour vision, but poor sense of smell. Therefore bird flowers are usually bright, with highly-saturated intense colours. In their systematic working of a spike or spray of flowers the honeyeaters usually take the nectar without damage to the flower, but may at times puncture the floral tube to get at the nectar.

The lorikeets are brush-tongued parrots which, in contrast to the majority of parrots which are seed-eater, have specialised in feeding upon the nectar and the pollen of flowering trees. The group includes some of the smallest of parrots, and has many extremely colourful members, such as the gaudy Rainbow Lorikeet.

Lorikeets are specialised for the nectar diet by the tongue, which is large and long, and can be pushed deep into the flower. Lorikeets appear to confine their attention to flowers of eucalypts, erythrinas and others which are not deep but are accessible to their short curved parrot beak. They open their beaks wide over the flowers and brush up nectar, tiny insects and pollen; for this purpose the tongue has a double brush along its upper surface.

The pollinating effect of lorikeets is probably limited to the brushing of the plumage of wings, tail and body over the flowers as they climb among the blossoms.

The flowers which by day attract these birds often are revisited at night by a variety of furred animals. The mammals that include nectar as part of their diet are the Honey Possum, Feathertail and Sugar Gliders, the Freckled Marsupial-mouse or Dibbler, and the blossom-bats.

The Honey Possum more than any other mammal has entered into a partnership with the flowering plants, and in the process has become very much modified for this way of life. The unique Honey Possum is the only species in its family; it is so different to any other mammal that it must be placed in a family of its own. Its body, snout and limbs have become so extremely modified that little is known of its ancestry except that it has no close relatives among the marsupials that inhabit Australia today, and it seems only distantly related to the similar sized pigmy possums.

The Honey Possum or Noolbenger is about the size of a small mouse, and has three well-defined stripes

Spinebill feeding at flower

down its back. The tail can be curled around twigs as a fifth hand, and its snout is very long and slender. It is found only in the south-west corner of Western Australia, where there is an abundance of large wildflowers throughout the year.

The most extreme adaptations for the nectar-eating way of life are to be seen in the modifications of the head, the slender snout being used like the beak of a honeyeating bird, to probe into narrow tubular flowers for nectar. Many of the most common wildflowers of this region are deeply tubular (grevilleas, lambertias) or brush-like (banksias, bottlebrushes) so the nectar can be reached only by an animal equipped to probe into the flowers, where the nectar might not be accessible to an ordinary short-tongued, blunt-faced mammal.

The Honey Possum's snout has become lengthened until now it is almost a trunk or proboscis. There are ridges along its lips which overlap to form a tubular channel through which nectar, pollen and microscopic insects (which live in great numbers in many of the flowers) can be sucked up. Its slender tongue is covered with fine bristles, with a tuft at the tip for licking up nectar, and darts out 25 mm beyond the tiny animal's nose. The teeth have very much degenerated after millions of years of soft food.

Honey Possums live in a region where, on the heathlands and banksia woodlands, there are at any time of the year some wildflowers in bloom, autumn and winter included. The little marsupials move about, possibly in colonies, according to the flowering, at one season being found in the bottlebrush scrub, and in another season, among the banksias.

Some flowers have evolved a unique partnership with certain nectar-eating bats; such flowers generally are large, strongly scented and coloured white or another pale colour easily seen at night.

In Australia two species of blossom bats occur in tropical northern and north-eastern Australia. Both are very small, and could easily be mistaken for ordinary insectivorous bats. They look rather like miniature flying foxes, with rather dog-like faces. Little is known of their habits except that they live largely upon nectar and pollen.

The Queensland Blossom Bat has a brush tongue like the honeyeating birds and the Honey Possum.

Some tropical Australian wildflowers have large, pale-coloured, strongly scented flowers, and it seems likely that these may be pollinated by bats. The tree *Sesbanfa formosa*, of the Kimberley region, has huge creamy-white flowers which hang in multiple racemes forming huge masses of flowers that must be easily found at night, and it could well be a bat pollinated species.

Flowers which have evolved for pollination by birds usually have strong stems and a stiff wiry flower structure as evident among the banksias with their brush-like texture. The New Holland Honeyeater, also called Yellow-winged Honeyeater, occurs in both eastern and western Australia, while this species of banksia grows in south-western Australia.

Right:

A male Western Spinebill, probing the flowers of a Scarlet Regelia shrub, will carry pollen daubed on the plumage around its face; pollen can be seen on the dark plumage of its forehead.

Below:

A Western Spinebill feeding at a grevillea shrub pushes its beak into the narrow tubular flower, pulling the long pollen-tipped style down on to its head. During spring and summer months each one of these honeyeaters carries a golden patch of pollen on its head. Entry to the nectary of these flowers appears restricted to birds whose sharp slender beaks can part the stiff, shell-like halves of the coiled flower tube, and spear through a barrier of woolly white hairs that blocks the tube just above the nectar.

Left:

The Sugar Glider, a small nocturnal gliding possum, feeds upon nectar and insects. Here it investigates the flowers of a bottlebrush. To get at the nectar it must thrust its face among the pollen-tipped crimson filaments, gathering pollen on its fur and spreading pollen among the flowers.

Below:

The Sugar Glider occurs in coastal forests of northern and eastern Australia from the Kimberleys to Adelaide. Here it is shown licking nectar from the flowers of a north Queensland eucalypt. Climbing among the flowers of eucalypts it would brush pollen from flower to flower, then, gliding to another tree as much as 46m away, carry pollen clinging to its whiskers and fur to other flowers, effectively bringing about the cross-pollination of the trees.

The unique Honey Possum has evolved a snout like the beak of a honeyeating bird. It is so specialised for its nectar diet that its teeth have become extremely small. Its lips and tongue form a tube, through which it can draw up nectar with its brush-tipped tongue.

8 The Rare Ones

From the beginning of white occupation of Australia, everything possible has been done to conquer and subdue the land, to obliterate the strange harsh Australian bush and replace it with the more familiar European type of environment. It has been estimated that within the first one hundred years of white settlement, one-third of the forests of New South Wales had been cut out to make way for farms and grazing lands with imported grasses for domesticated livestock. During the same period some 165 kinds of foreign plants were introduced, and spread rapidly. European animals were liberated both accidentally and deliberately. Soon many of these had become serious pests, like the prickly pear which took over sixty million acres of land, and the rabbit which spread from coast to coast in enormous numbers.

Innumerable other weeds and pests have become widely established and are to be found almost anywhere in the Australian bush, the foreign plants replacing the natives, the introduced animals competing with or preying upon the native fauna.

Equally disastrous has been the effect of flocks of sheep and herds of cattle. Across vast areas of inland Australia the defenceless small native animals are disappearing as stock destroy much, in places all, of the natural scrub, the ground cover needed to shelter the little native mammals.

Much of the inland, originally well vegetated with tough stunted scrub, lavish in its wildflower displays and rich in wildlife in good seasons, is in danger of becoming a real desert. Already big areas of man-made desert can be found, where the natural scrub has gone through over-grazing.

The native animals of these areas have suffered more losses than anywhere else in Australia. The dry inland plains are a fragile environment, more vulnerable to steady, almost un-noticed deterioration of habitat, than are the more fertile coastal regions.

The trampling action of the hooves of sheep and cattle has compacted the soil. No longer could animals such as the bandicoots dig their tunnels or scratch around for insect larvae, which in any case would be gone. Animals such as the beautiful Rabbit-eared Bandicoot therefore need large reserves where domestic stock are not allowed.

Each of our unique native animals is the culmination of some six hundred million years of evolution, interwoven with the evolution of all life on our planet, including the shaping of our own species over the ages.

But it is already too late for many species of unique Australian wildlife. Some exist only as bones and skins in museum cabinets. Others are so perilously close to extinction that it may not be in our power to save them. Some have lost the natural habitat without which they cannot survive; with others the populations are so small that the basis of a permanent breeding colony may not exist. But with most of our rare fauna we really do not know exactly how rare they are. Some may be, like the Leadbeater's Possum of Victoria, reasonably abundant, yet never seen for long periods of time; others, like the tiny Kimberley Planigale, are probably rare in number of specimens collected simply because their habitat is remote and largely inaccessible.

Many have been unsighted so long that they have seemed extinct, yet have been rediscovered, and undoubtedly more will be found. Most of the rediscoveries have concerned small nocturnal animals in areas where there is heavy forest or other thick vegetation. But some of the missing animals lived in areas that now are nothing but bare paddocks, roads and barb-wire fences. For these there can be little hope.

One of the first and most publicised of the series of exciting rediscoveries of recent years was the finding of the Noisy Scrub-bird.

The disappearance of the Noisy Scrub-bird was one of Australia's greatest ornithological mysteries. Discovered in 1843 by John Gilbert, field collector for the famous English ornithologist John Gould.

In eastern Australia there exists a second species of scrub-bird, the Rufous, which inhabits the mountain-top rainforests of north-eastern New South Wales and south-eastern Queensland.

Brush-tailed Bettong

These two species of scrub-bird are thought to be the most primitive. Unlike more advanced song-birds, they lack any wishbone, and have a more basic, prototype arrangement of the syrinx or voice-box muscles.

About a dozen specimens of the Noisy Scrub-bird were collected in various parts of the south-western corner of the continent, but none after 1889. From that time onwards there was neither sight nor sound of the Noisy Scrub-bird in any of its known haunts.

As the decades passed the disappearance of this scientifically interesting bird became more and more of a mystery, and the 'extinct' bird became almost a legend among ornithologists. Then quite accidently and unexpectedly the species was rediscovered in 1961.

In December of that year a school headmaster, Harley Webster, was fishing at Two People Bay, on the south coast of Western Australia about 32 km east of Albany. The path to the beach here passes through dense swampy paperbark and reedy vegetation around a little creek, and it was here that he heard a bird calling in a voice of extraordinary loudness.

Since then careful searching of all likely places has shown that there is a small population of perhaps forty pairs of birds surviving in pockets of low, dense vegetation in gullies and hollows on the slopes of Mt Gardner, overlooking Two People Bay. This hill and the surrounding coastal heathlands now form a fauna reserve of about 4,850 ha, under the careful supervision of a full-time ranger.

Except for some catastrophic event such as a big bushfire, the Noisy Scrub-bird seems secure for the present time.

Among the mammals of Australia there has been an encouraging series of rediscoveries. First was the finding of Leadbeater's Possum near Melbourne in 1961. This beautiful little marsupial is about 30 cm long, nose to tip of the long bushy tail; it has a stripe on the face, and looks rather like the common Sugar Glider but lacks the gliding membranes.

Very few specimens of Leadbeater's Possum had ever been collected, and all were taken around 1867 when it was first described and named. In spite of intensive searches in the localities of the original specimens, no trace of the little marsupial could be found, so after more than ninety years it seemed certainly extinct. One fact alone offered hope — the known habitats were in extremely mountainous country, with dense forests. A small nocturnal could possibly go un-noticed, even for as long as ninety years.

Leadbeater's Possum was rediscovered by a party of naturalists using spotlights for a survey of the nocturnal mammals of the Cumberland Valley, only 113 km from Melbourne. It was glimpsed for a moment climbing in the tree tops, showing itself just long enough to be distinguished from the common Sugar Glider by the difference in its tail. Once the possum's habitat was known, naturalists found the possum in many similar places through the ranges, and it is now

Noisy Scrub-Bird

considered to be quite a common species, though restricted in distribution.

Two other rediscoveries of recent years have been the Parma Wallaby, and a small carnivorous marsupial-mouse known as the Dibbler.

It seems that the Parma was never a very common species, even at the time of first settlement, when it lived in the rainforests around Lake Illawarra, New South Wales. The last record of the Parma was in 1932, when two were collected near Dorrigo, 480 km further up the coast.

This attractive little wallaby was rediscovered in 1965, as a result of investigations made by Dr W.D.L. Ride, Director of the Western Australian Museum. Ride examined wallaby skins sent from New Zealand, and thought some of these could be the Parma. Further investigation showed that small wallabies that lived in great numbers on Kawau Island, off the coast of New Zealand, were indeed Parmas. They had been taken there very many years ago by Sir George Grey, a governor of New Zealand, and once an Australian explorer with a deep interest in Australian animal life.

Not only did the extinct Parma exist in New Zealand, but in such great numbers that Parmas were being shot by the thousand because they were damaging pine plantations. Small numbers have now been returned to their homeland, as the beginning of a re-establishment programme. By a remarkable chance, the Parma has since been discovered to survive still in forest country near Gosford.

The rediscovery of the Dibbler, or Freckled Marsupial-mouse, on the south coast of Western Australia east from Albany, was quite accidental and totally unexpected. It shows how easily a small marsupial can live without being noticed, in an area frequented by fishermen, tourists and occasional zoologists, near farms and not far from towns.

This rare marsupial was rediscovered when a female was trapped on flowers of *Banksia* trees, by the author, in January 1967. It was the first specimen since 1884. A male was captured in similar fashion two nights later.

The rediscovery of the Dibbler was made quite accidently during an attempt to find a specimen of the Honey Possum, which occurs in that locality. As the Honey Possum was known to be a nectar feeder, cage traps were specially constructed to fit over the large flower heads, with a thread-like tripwire to spring the door shut passing across the surface of the flower. But instead of the Honey Possum, it was the far more exciting and rare Dibbler that tripped the door shut. A third specimen was caught several months later.

Since 1967 attempts by other naturalists to locate further living specimens have been unsuccessful, so the Dibbler remains one of our rarest mammals, with only three individuals being seen in eighty-three years, all coming from the 200 m^2 area of low coastal sand-heath scrub on the lower slopes of Mt Manypeaks, on the south coast of Western Australia.

Leadbeaters Possum

Left:

One of the rarest of wallabies in Australia, the Parma or White-fronted Wallaby was thought to be extinct. It was rediscovered not in its native land, but in New Zealand, on the small island of Kawau, where it had been introduced, and then forgotten. Small numbers have been returned to Australia.

Below:

The Dibbler remains one of Australia's rarest mammals, only three having been caught in eighty-three years. Probably a small population still survives in an area of dense low coastal undergrowth on the south coast of Western Australia. Until its rediscovery in 1967, it had not been seen since 1884.

Rarely seen, and photographed on two or three occasions only, the Rufous Scrub-bird has a distribution restricted to the mountain-top rainforests of south-eastern Queensland and north-eastern New South Wales. An accomplished mimic and ventriloquist, the male calls loudly from dense vegetation on the floor of the rainforest. He keeps on the ground much of the time, and, running swiftly through the tangled masses of ferns, vines and other jungle vegetation, can be very difficult to see even when calling close by. A very similar and even more rare species, the Noisy Scrub-bird, inhabits coastal thickets in south-western Australia.

The Southern Potoroo is one of the little rat-kangaroos which have become so rare in most parts of Australia following the destruction of much of the dense vegetation which is essential for their protection and shelter.

Ground Parrot

BIRDS AND MAMMALS BELIEVED TO BE EXTINCT

MARSUPIALS:

Broad-faced Potoroo *Potorous platyops* (W.A.)
Brown or Eastern Hare-wallaby *Lagorchestes leporides* (N.S.W., Vic., S.A.)
Toolache Wallaby *Macropus greyi* (Vic., S.A.)
Moonie River Wombat *Lasiorhinus gillespiei* (Qld)

PLACENTALS:

Short-tailed Hopping-mouse *Notomys amplus* (N.T.)
Darling Downs Hopping-mouse *Notomys mordax* (Qld)
Long-tailed Hopping-mouse *Notomys longicaudatus* (W.A.)
White-tipped Stick-nest Rat *Leporillus apicalis* (Central Aust.)
White-footed Tree-rat *Conilurus albipes* (Qld, N.S.W., S.A.)

BIRDS:

Tasmanian Emu, a sub-species of the common Emu *Dromaius novae-hollandiae* (Tas.)
King Island Emu, another sub-species of the common Emu (King Island, Bass Strait)
Paradise Parrot *Psephotus pulcherrimus*, not seen since 1936. (Qld, N.S.W.)
Night Parrot *Geopsittacus occidentalis*, last seen 1931 (inland Australia)

BIRDS AND MAMMALS RARE OR ENDANGERED

MARSUPIALS:

Parma Wallaby *Macropus parma* (N.S.W.)
Crescent Nail-tailed Wallaby *Onychogalea lunata* (southern Aust.)
Bridled Nail-tailed Wallaby *O. fraenata* (N.S.W., Qld)
Central Australian Hare-wallaby *Lagorchestes asomatus* (N.T., W.A.)
Banded Hare-wallaby *Lagostrophus fasciatus* (W.A.)
Desert Rat-kangaroo *Caloprymnus campestris* (S.A., Qld)
Boodie, or Lesueur's Rat-kangaroo *Bettongia lesueur* (W.A., S.A., N.T., N.S.W.)
Brush-tailed Bettong or Woilie *Bettongia penicillata* (W.A., S.A., Vic., N.S.W.)
Queensland Hairy-nosed Wombat *Lasiorhinus barnardi* (Qld, N.S.W.)
Scaly-tailed Possum *Wyulda squamicaudata* (W.A.)
Leadbeater's Possum *Gymnobelideus leadbeateri* (Vic.)
Mountain Pigmy Possum *Burramys parvus* (Vic.)
Pig-footed Bandicoot *Chaeropus ecaudatus* (N.S.W., Vic., S.A., N.T., W.A.)
Bilby, Dalgyte or Common Rabbit-eared Bandicoot *Macrotis lagotis* (W.A., S.A., N.T., Qld)
Lesser Rabbit-eared Bandicoot *Macrotis leucura* (Central Aust.)
Spiny Bandicoot *Echymipera rufescens* (Qld)
Desert Bandicoot *Perameles eremiana* (S.A., N.T., W.A.)
Tasmanian Wolf or Tiger *Thylacinus cynocephalus* (Tas.)
Kultarr *Antechinomys laniger* (Vic., N.S.W., Qld)
Sandhill Dunnart *Sminthopsis psammophila* (N.T.)
Long-tailed Dunnart *Sminthopsis longicaudata* (W.A., N.T.)
White-tailed Dunnart *Sminthopsis granulipes* (W.A.)
Red-tailed Phascogale, Red-tailed Wambenger, *Phascogale calura* (W.A., N.T., S.A., Vic., N.S.W.)
Narrow-nosed Planigale *Planigale tenuirostris* (N.S.W., W.A., Qld)
Kimberley Planigale *Planigale subtilissima* (W.A.)
Dibbler or Freckled Marsupial-mouse *Antechinus apicalis* (W.A.)
Woodward's Wallaroo *Macropus bernardus* (N.T.)
Northern Rat-kangaroo *Bettongia tropica* (Qld)
Rock-haunting Ringtail *Petropseudes dahli* (N.T., W.A.)
Long-tailed Pigmy Possum *Cercartetus caudatus* (Qld)

PLACENTALS:

False Swamp-rat *Xeromys myoides* (Qld, N.T.)
Big-eared Hopping-mouse *Notomys megalotis* (W.A.)
Short-tailed Hopping-mouse *Notomys amplus* (N.T.)
Macdonnell Range Rock-rat *Zyzomys pedunculatus* (N.T.)
Hastings River Mouse *Pseudomys oralis* (N.T.)
New Holland Mouse *Pseudomys novaehollandiae* (N.S.W.)
Smokey Mouse *Pseudomys fumeus* (Vic.)
Western Mouse *Pseudomys occidentalis* (W.A.)
Little Northern Territory Long-eared Bat *Nyctophilus walkeri* (N.T.)
Warty-nosed Horseshoe Bat *Hipposideros semoni* (Qld)
Lesser Warty-nosed Horseshoe Bat *Hipposideros stenotis* (N.T., Qld)
Norfolk Island Scurrying Bat *Tadarida norfolkensis* (Norfolk Is.)
Small-footed Myotis *Myotis australis* (N.S.W.)
Hughenden Broad-nosed Bat *Nycticeius influatus* (Qld)

BIRDS:

Freckled Duck *Stictonetta naevosa* (throughout Aust.)
Golden-shouldered Parrot *Psephotus chrysopterygius* (Qld)
Hooded Parrot, a sub-species of the Golden-shouldered (N.T.)
Orange-bellied Parrot *Neophema chrysogaster* (Tas., Vic., S.A.)
Turquoise Parrot *Neophema pulchella* (Qld, N.S.W., Vic.)
Scarlet-chested Parrot *Neophema splendida* (W.A., S.A.)
Noisy Scrub-bird *Atrichornis clamosus* (W.A.)
Western Whipbird *Psophodes nigrogularis* (W.A., S.A., Vic.)
Eyrean Grass-wren *Amytornis goyderi* (S.A.)
Helmeted Honeyeater *Meliphaga cassidix* (Vic.)

Index

ITALICISED NUMERALS REFER TO ILLUSTRATIONS